MW01199400

EINSTEIN
ON COSMIC RELIGION
AND OTHER OPINIONS AND APHORISMS

ALBERT EINSTEIN

With an Appreciation by
GEORGE BERNARD SHAW

DOVER PUBLICATIONS, INC.
Mineola, New York

Bibliographical Note

This Dover edition, first published in 2009, is an unabridged republication of *Cosmic Religion with other Opinions and Aphorisms,* originally published in 1931 by Covici-Friede, Inc., New York.

Library of Congress Cataloging-in-Publication Data

Einstein, Albert, 1879–1955.
Einstein on cosmic religion : and other opinions and aphorisms / Albert Einstein ; with an appreciation by George Bernard Shaw.
p. cm.
ISBN-13: 978-0-486-47010-8
ISBN-10: 0-486-47010-5

2008051832

Manufactured in the United States by Courier Corporation
47010502
www.doverpublications.com

GRATEFUL acknowledgment is hereby made to the *New York Times Magazine* for *Cosmic Religion*; to the Jewish Telegraphic Agency for material from its files; and to the New History Society for *Militant Pacifism*.

CONTENTS

BIOGRAPHICAL NOTE

BIOGRAPHICAL NOTE

The following sketch has been prepared by the publishers, in the belief that it may prove of aid to the readers of this book, and is based, in part, on material prepared by Dr. Einstein's representatives, and on various published articles and books.

Albert Einstein, discoverer of the theory of relativity, and one of the greatest thinkers of the modern era, was born at Ulm, on the Danube, March 14th, 1879. His early years were spent in Munich where his father had become part owner of an electro-technical plant, and it was here that the boy, who, curiously enough, showed no extraordinary aptitude or brilliance as a student, nevertheless received the initial impulses which turned his thoughts to physics. His childhood was lonely; the mechanical regulations of the schools and their rigid, unimaginative enforcement—the inevitable concomitants of the German education of his day—were a severe trial to his sensitive spirit and he found refuge for his inarticulate longings in a profound religious feeling which he soon began to express in brief songs. After several years, the decline of his father's business compelled the family to move to Milan, and their comfortable, middle-class life came to an end. Einstein was by then 15 years old and his period of care began.

In Milan he left school for a while to study art. Italy was more congenial to his spirit than his native Germany—the curious affinity of intellectual Germans for the warmer Italian sky is here again demonstrated—and his genius began to develop. The necessity for earning a living sent him to Zurich to enter the Polytechnic Academy. He failed in his first attempt but a year later, after a course at the Canton School of Aarau, he succeeded in entering and soon displayed outstanding ability in mathematics and physics. He remained in Switzerland for several years, following his student years as an instructor at Zurich and in 1902 assuming a post in the Bern Patent Office where he served as engineer and technical advisor. His free time he devoted to the pursuit of his own scientific speculations.

In 1905 the event whose twenty-fifth

anniversary has just been signalized in world-wide demonstrations and celebrations burst upon an unprepared and skeptical scientific world. This event was the publication in the *Annals of Physics* (*Annalen der Physik*) of Einstein's first statement of what later was to become famous as the *Special Theory of Relativity*. He was then 26 years old, and the scientists of the day laughed at the obscure heretic who thus boldly challenged the Newtonian law of gravitation and the physical universe postulated by the great Englishman. That first essay was called *On the Electro-Dynamics of Moving Bodies* and it has been the basis for his life-work. All that Einstein has since done has been in the nature of expansion and clarification of that initial statement.

He became assistant professor at the Zurich Academy in 1909. In 1911 he was offered and accepted a professorship at Prague, but two years later he returned to Zurich as Professor in Theoretical Physics at the Polytechnic Academy where he had been educated. Some months prior to the outbreak of the World War he was called to the Prussian Academy of Sciences at Berlin and was appointed Director of the newly-founded Kaiser Wilhelm Institute for Physics. With this appointment came the opportunity to arrange and formulate the daring speculations which had occupied his thought during the preceding eight years and he now began the preparation of his great work, finally published in 1916, as *The General Theory of Relativity*. This work he has since subjected to several modifications as the lines of his own investigations suggested, and in 1929 he announced his *Unitary Field Theory*, which is an attempt to reduce the laws of gravitation and electro-magnetics to a unified system of coordinated mathematics. This fusion he achieves by the use of a new kind of space-time geometry specially designed for the purpose. Recently, in various addresses, he has announced that his next step would be to bring the submicroscopic world of electrons and protons, as well as quanta, into the fold of one unified mathematical law. And he has also stated his new hypothesis of space as the one, all-embracing reality.

* * *

This bare account outlines the life-work of the most original and daring speculative thinker and scientific investigator since Newton. Einstein's fame rests on his two published theories and on the astronomical observations of physical phenomena which confirmed them. He has overthrown a universe which endured for three centuries and in its place has constructed a new one, incapable of comprehension to man's senses but subject to imprisonment and expression within the symbols of Einstein's mathematical formulas. He has destroyed the hitherto existing foundations of all physical science; he has added another dimension to the universe—time; and he has introduced us to this new universe which, perhaps within the span of another four or five generations, will become as commonplace to the ordinary man as is the present one.

Save in such elementary examples by which the newspaper public has been instructed, no simple, general statement of Einstein's theories is possible and no attempt to define it need here be made. For the reader who is interested, there are several good analyses in English which meet the demands of the non-technical, lay student. What is important to mention, however, is the fact that the central idea of his work is a problem which has long occupied the minds of philosophers. Relativity, in itself, is not a new idea. The problem has been recognized for generations and the mind of man has long wrestled with its implications. But there is a vast difference between the philosopher who says "My idea of the world is the real world. But my neighbor has his own idea of the world and to him his idea is equally real. Hence there is no reality, only an idea of reality which each man makes for himself," and the theory propounded by Einstein which for the first time gives scientific validity to this conception by establishing its mathematical proofs. This is perhaps the clearest demonstration of the fact that Einstein is the imaginative thinker first and the scientist second. The scientist works in observatories and laboratories and from the observed results of thousands of experiments announces a deduced general law which accounts for the phenomena recorded. Einstein, the intuitive thinker, conceives an idea, develops its mathematical soundness, and then establishes its truth on the basis of observed phenomena. This is the reverse of the empirical scientific method and it is for this reason, perhaps, that his earliest statement of his ideas met with incredulity and ridicule.

Well, his case has now been won. To the world-old search for a 'standard of reference' by which to gauge the workings of the physical universe Einstein retorts that there is no standard of reference. A yard-stick is one yard long only while it is at rest. Move it and it becomes shorter, and the faster it is moved the shorter it becomes. Similarly, we are accustomed to measure the events of our world in terms of time, the elapsing of which we consider an unalterable, absolute measure. Einstein dethrones this fetich of absolute time. His exact analysis of the idea of simultaneity demonstrates that every body in motion has its own time which elapses more slowly as the body moves more rapidly. Thus, the watch of the pedestrian runs faster than that of the flyer. An hour in the coffee-house is shorter than an hour in the subway: we have no right to compare two differently moved clocks; not that the subway clocks falls behind but that the time of the running train elapses more slowly. Because of the slight speeds involved in these examples, the differences in time are infinitesimal. However, when a body has a light-speed (186,000 miles per second), its 'time' ceases altogether, its 'clock' stands still. Einstein has said, "If a person were hurled at the velocity of light away from the earth and from a certain point allowed to return at the same speed, he would not become a second older in the interim even though the time of the earth had elapsed a thousand years while he was on his journey."

Biographical Note

Our physical universe as we have conceived it crumbles in the light of these teachings. Copernicus destroyed the notion of the absolute repose of the earth; Einstein destroys Absolutism entirely. Nothing is absolute, all is relative. There is thus no reality—only the Einsteinian reality of space. We do not live in a three-dimensional world of length, breadth, and height, in which temporal changes take place, but in a four-dimensional universe whose equally entitled dimensions are length, breadth, height, and time.

How long Einstein's theories will continue to hold good cannot, of course, be predicted. He has himself made important modifications in them during the last twenty-five years, and his latest work, as yet not finally or definitely stated, does not carry to physicists the same inevitable logic and self-consistency which marked the earlier theories. Einstein is himself aware of these weaknesses and probably these points can never be resolved until gravitational and atomic theory have been brought into closer relationship. But as far as they go they have altered the entire direction of physical investigation and the energies of the world's scientists for generations will stem from the work of this German professor who is at once the most imaginative thinker and the greatest scientist of modern times.

* * *

Biographical Note

Einstein the man is a much simpler person than his fame would lead one to suppose. He derives his greatest enjoyment from his piano, on which he improvises, and from his violin, on which he plays with the skill of a virtuoso. He is an ardent yachtsman, is fond of long, aimless tramps in the country (recall Nietzsche and his dictum that only thoughts conceived while walking are worth while), is deeply interested in modern world affairs, and has lent the enormous weight of his reputation to the service of his race in its struggle for national autonomy and freedom from oppression. He is what he himself describes as a 'militant pacifist' and unalterably opposed to war on any grounds. He hates crowds as much as he does honors, ceremonials, and speeches. To a man of his ironic mind, the acclaim of a multitude incapable of understanding ten words of his published work must be singularly amusing.

Nevertheless, the world has insisted on according its recognition to his genius. He won the Nobel Prize for Physics in 1921 and in 1926 the Royal Astronomical Society of London presented him with its gold medal. He has been honored by innumerable scientific and educational institutions and has received the homage of the most distinguished and prominent men of his time. His first long tour, in 1921, which led him to America, was in the nature of a triumphal procession and has been eclipsed only by the welcome which greeted him on his second visit to the United States in 1930. He stands unique and supreme among his contemporaries—the greatest mind and one of the most lovable men alive.

A whole literature has grown up around the Einstein theories and the complete bibliography now totals more than 4000 separate works. Of these, a few of the useful books and articles in English are:

EINSTEIN, ALBERT: RELATIVITY; THE SPECIAL AND GENERAL THEORY. New York, 1920.

—— SIDELIGHTS ON RELATIVITY. London, 1922.

—— THE MEANING OF RELATIVITY. Princeton, 1923.

—— THE UNITARY FIELD THEORY. Brooklyn, N. Y., 1929.

—— ABOUT ZIONISM. London, 1930.

BROSE, HENRY L.: THE THEORY OF RELATIVITY. Oxford, 1920.

COLLIER, JOHN: EINSTEIN FOR THE NON-MATHEMATICAL. The *National Review*, vol. 94, pp. 123-125. London, 1929.

EDDINGTON, A. S.: EINSTEIN'S THEORY OF SPACE AND TIME. The *Contemporary Review*, vol. 116, pp. 639-643. New York, 1919.

GUGGENHEIMER, SAMUEL H.: THE EINSTEIN THEORY EXPLAINED AND ANALYZED. New York, 1925.

MOSZKOWSKI, ALEXANDER: EINSTEIN, THE SEARCHER. London, 1921.

NORDMANN, CHARLES: EINSTEIN AND THE UNIVERSE. New York, 1922.

REISER, ANTON: ALBERT EINSTEIN. New York, 1930.

SLOSSON, EDWIN E.: EASY LESSONS IN EINSTEIN. New York, 1921.

AN APPRECIATION BY
GEORGE BERNARD SHAW

AN APPRECIATION

BY GEORGE BERNARD SHAW *

. . . Napoleon and other great men were makers of empires, but these eight men whom I am about to mention were makers of universes and their hands were not stained with the blood of their fellow men. I go back 2,500 years and how many can I count in that period? I can count them on the fingers of my two hands. Pythagoras, Ptolemy, Kepler, Copernicus, Aristotle, Galileo, Newton, and Einstein, and I still have two fingers left vacant.

* From an address delivered at the dinner for Professor Einstein in London, October 27, 1930.

Since the development of Newton 300 years ago there have been nine generations of men, and those nine generations of men have not enjoyed the privileges that we are enjoying here to-night. We are standing face to face with one of those great men, looking forward to the privilege of hearing his voice, and maybe another 300 years will pass before another generation will enjoy that privilege. Even among those eight men I must make a distinction. I have called them makers of the universe, but some of them were only repairers. Only three of them made universes. Newton made a universe which lasted for 300 years. Einstein has made a universe, which I suppose you want me to say will never stop, but I don't know how long it will last.

These great men, they have been the makers of one side of humanity, which has two sides. We call the one side religion, and we call the other science. Religion is always right. Religion protects us against that great problem which we all must face. Science is always wrong; it is the very artifice of men. Science can never solve one problem without raising ten more problems.

What have all of these great men been doing? Each in turn claimed the other was wrong, and now you are expecting me to say that Einstein proved that Newton was wrong. But you forget that when science reached Newton, science came up against that extraordinary Englishman. That had never happened to it before. Newton lent a power so extraordinary that if I was speaking fifteen years ago, as I am old enough to have done, I would have said that he had the greatest mind that ever man was endowed with. Combine the light of that wonderful mind with credulity, with superstition. He knew his people, he knew his language, he knew his own folk, he knew a lot of things; he knew that an honest bargain was a square deal and an honest man was one who gave a square deal. He knew his universe; he knew that it consisted of heavenly bodies that were in motion, and he also knew the one thing you cannot do to anything whatsoever is to make it move in a straight line. In other words, motion will not go in a straight line.

If you take a poor man and blindfold that man and say, "I will give you a thousand pounds if you, blindfolded, will walk in a straight line," he will do his best for the sake of the thousand pounds to walk in a straight line, but he will walk in a circle and come back in exactly the same place.

Mere fact will never stop an Englishman. Newton invented a straight line, and that was the law of gravitation, and when he had invented this, he had created a universe which was wonderful in itself. When applying his wonderful genius, when he had completed a book of that universe, what sort of book was it? It was a book which told you the station of all the heavenly bodies. It showed the rate at which they were traveling; it showed the exact hour at which they would arrive at such and such a point to make an eclipse. It was not a magical, marvelous thing; it was a matter-of-fact thing, like a Bradshaw.

For 300 years we believed in that Bradshaw and in that Newtonian universe as I suppose no system has ever been believed in before. I know I was educated in it and was brought up to believe in it firmly. Then a young professor came along. He said a lot of things and we called him a blasphemer. He claimed Newton's theory of the apple was wrong.

He said: "Newton did not know what happened to the apple, and I can prove this when the next eclipse comes."

We said: "The next thing you will be doing is questioning the law of gravitation."

The young professor said: "No, I mean no harm to the law of gravitation, but, for my part, I can go without it."

"What, do you mean, go without it?"

He said: "I can tell you about that afterward."

The world is not a rectilinear world; it is a curvelinear world. The heavenly bodies go in curves because that is the natural way for them to go, and so the whole Newtonian universe crumpled up and was succeeded by the Einstein universe. I am sorry to have to say it. You must remember that our distinguished visitor could not have said it. It would not be nice for him to say it; it would not be courteous. But here in England is a wonderful man. This man is not challenging the fact of science; he is challenging the action of science. Not only is he challenging the action of science, but the action of science has surrendered to his challenge.

When Newton said the line of nature is a straight line, William Hogarth said the line of nature is a curve. He anticipated our guest.

I have talked enough. I rejoice in the new universe that Einstein has produced.

An Appreciation

This is a very distinguished assembly, but it is not an assembly composed exclusively of Einsteins. With a genius a certain solitude is inevitable. I will ask him to remember this, that in our human way we all have our little solitudes. My friend Mr. Wells has spoken to us of the secret sessions of the heart. Our lives are so small that we are too often in our solitude like children crying in the dark. Nevertheless our little solitude is a great and august solitude in which we can contemplate things that are greater than mankind.

COSMIC RELIGION

COSMIC RELIGION

Everything that men do or think concerns the satisfaction of the needs they feel or the escape from pain. This must be kept in mind when we seek to understand spiritual or intellectual movements and the way in which they develop. For feeling and longing are the motive forces of all human striving and productivity—however nobly these latter may display themselves to us.

What, then, are the feelings and the needs which have brought mankind to religious thought and to faith in the widest sense? A moment's consideration shows that the most varied emotions stand at the cradle of religious thought and experience.

In primitive peoples it is, first of all, fear that awakens religious ideas—fear of hunger, of wild animals, of illness, and of death. Since the understanding of causal connections is usually limited on this level of existence, the human soul forges a being, more or less like itself, on whose will and activities depend the experiences which it fears. One hopes to win the favor of this being by deeds and sacrifices, which, according to the tradition of the race, are supposed to appease the being or to make him well disposed to man. I call this the religion of fear.

This religion is considerably stabilized—

though not caused—by the formation of a priestly caste which claims to mediate between the people and the being they fear, and so attains a position of power. Often a leader or despot, or a privileged class whose power is maintained in other ways, will combine the function of the priesthood with its own temporal rule for the sake of great security; or an alliance may exist between the interests of the political power and the priestly caste.

* * *

A second source of religious development is found in the social feelings.

Fathers and mothers, as well as leaders of great human communities, are fallible and mortal. The longing for guidance, for love and succor, provides the stimulus for the growth of a social or moral conception of God. This is the God of Providence, who protects, decides, rewards, and punishes. This is the God who, according to man's widening horizon, loves and provides for the life of the race, or of mankind, or who even loves life itself. He is the comforter in unhappiness and in unsatisfied longing, the protector of the souls of the dead. This is the social or moral idea of God.

It is easy to follow in the sacred writings of the Jewish people the development of the religion of fear into the moral religion, which is carried further in the New Testament. The religions of all the civilized peoples, especially those of the Orient, are principally moral religions. An important advance in the life of a people is the transformation of the religion of fear into the moral religion. But one must avoid the prejudice that regards the religions of primitive peoples as pure fear religions and those of the civilized races as pure moral religions. All are mixed forms, though the moral element predominates in the higher levels of social life. Common to all these types is the anthropomorphic character of the idea of God.

Only exceptionally gifted individuals or especially noble communities rise *essentially* above this level; in these there is found a third level of religious experience, even if it is seldom found in a pure form.

I will call it the cosmic religious sense. This is hard to make clear to those who do not experience it, since it does not involve an anthropomorphic idea of God; the individual feels the vanity of human desires and aims, and the nobility and marvelous order which are revealed in nature and in the world of thought. He feels the individual destiny as an imprisonment and seeks to experience the totality of existence as a unity full of significance. Indications of this cosmic religious sense can be found even on earlier levels of development—for example, in the Psalms of David and in the Prophets. The cosmic element is much stronger in Buddhism, as, in particular,

Schopenhauer's magnificent essays have shown us.

The religious geniuses of all times have been distinguished by this cosmic religious sense, which recognizes neither dogmas nor God made in man's image. Consequently there cannot be a church whose chief doctrines are based on the cosmic religious experience. It comes about, therefore, that precisely among the heretics of all ages we find men who were inspired by this highest religious experience; often they appeared to their contemporaries as atheists, but sometimes also as saints. Viewed from this angle, men like Democritus, Francis of Assisi, and Spinoza are near to one another.

How can this cosmic religious experience

be communicated from man to man, if it cannot lead to a definite conception of God or to a theology? It seems to me that the most important function of art and of science is to arouse and keep alive this feeling in those who are receptive.

Thus we reach an interpretation of the relation of science to religion which is very different from the customary view. From the study of history, one is inclined to regard religion and science as irreconcilable antagonists, and this for a reason that is very easily seen. For any one who is pervaded with the sense of causal law in all that happens, who accepts in real earnest the assumption of causality, the idea of a Being who interferes with the sequence of events in the world is absolutely impossible. Neither the religion of fear nor the social-moral religion can have any hold on him. A God who rewards and punishes is for him unthinkable, because man acts in accordance with an inner and outer necessity, and would, in the eyes of God, be as little responsible as an inanimate object is for the movements which it makes.

* * *

Science, in consequence, has been accused of undermining morals—but wrongly. The ethical behavior of man is better based on sympathy, education, and social relationships, and requires no support from religion. Man's plight would, indeed, be sad if he had to be kept in order through fear of punishment and hope of rewards after death.

It is, therefore, quite natural that the churches have always fought against science and have persecuted its supporters. But, on the other hand, I assert that the cosmic religious experience is the strongest and the noblest driving force behind scientific research. No one who does not appreciate the terrific exertions, and, above all, the devotion without which pioneer creations in scientific thought cannot come into being, can judge the strength of the feeling out of which alone such work, turned away as it is from immediate practical life, can grow. What a deep faith in the rationality of the structure of the world and what a longing to understand even a small glimpse of the reason revealed in the world there must have been in Kepler and Newton to enable them to unravel the mechanism of the heavens, in long years of lonely work!

Anyone who only knows scientific research in its practical applications may easily come to a wrong interpretation of the state of mind of the men who, surrounded by skeptical contemporaries, have shown the way to kindred spirits scattered over all countries in all centuries. Only those who have dedicated their lives to similar ends can have a living conception of the inspiration which gave these men the power to remain loyal to their purpose in spite of countless failures. It is the cosmic religious sense which grants this power.

A contemporary has rightly said that the only deeply religious people of our largely materialistic age are the earnest men of research.

PACIFISM

MILITANT PACIFISM

One of the problems of pacifism is, that when pacifists come together, they usually have the feeling that they are consorting with the sheep while the wolves are outside. Thus they reach only their own kind who are already convinced, and do not advance very far. That is the weakness of the pacifist movement.

The real pacifists, those who are not up in the clouds, but who think and count realities, must give up idle words, and fearlessly try to accomplish something of definite value to their cause.

We all know that when a war comes, every man accepts the duty to commit a crime—the crime of killing—each man for his own country.

Now those who realize the immorality of war should do their utmost to disentangle themselves from this old idea of military duty—and so become liberated from slavery. And for this liberation I have two suggestions: The first has, during war times, been tried and practiced in the past, by those who at great personal sacrifice have refused to do war service. However, the sincere pacifists to-day who mean to accomplish something must take this stand in times of peace, and in those countries where military service is compulsory the effect will be great. On the other hand, in other countries where military service is not compulsory, these same pacifists should openly assert that in case of war, they themselves would not participate. I recommend the recruiting of people with this idea in all parts of the world. And to the timid ones who fear imprisonment by their governments I say: "You need not fear imprisonment, for if you get only two per cent of the population of the world to declare in times of peace, 'We are not going to fight; we need other methods to settle international disputes,' this two per cent will be sufficient—for there are not jails enough in the world to hold them!"

The second method which I suggest

appears less illegal. I believe that international legislation should be advocated to the effect that those who declare themselves as war resisters should be allowed during peace times to take up different kinds of strenuous or even dangerous work, either for their own countries or for the international benefit of mankind. This would prove that they do not oppose war for their own private comfort or because they are cowards or because they do not want to serve their own country or humanity.

If, in order to prove this, we burden ourselves with these various strenuous and dangerous occupations, we shall have gone far toward achieving the pacification of the world. I am convinced that such legislation can be brought about.

I suggest to your organization* that you discuss these proposals at your coming meetings and adopt them, and I am pretty sure that whosoever takes the initiative along these lines, will, sooner or later, bring about such international legislation.

I further suggest that war resisters should organize themselves internationally and collect funds to support those resisters in the different countries who to-day cannot make progress because of lack of financial backing.

I advise and advocate very warmly and strongly the creation of an International

* The New History Society.

War Resisters' Fund to support the active war resisters of our day.

My final word to you is that those who are ambitious and sincerely dedicated to the cause of universal peace must have the courage to start, to initiate, and to carry on so fearlessly that the whole world will be forced to consider what they are doing!

DISARMAMENT

It has not become a generally recognized axiom that the giant armaments of all nations are proving highly injurious to them collectively.

I am even inclined to go a step further by the assertion that, under present day conditions any one state would incur no appreciable risk by undertaking to disarm—wholly regardless of the attitude of the other states.

If such were not the case it would be quite evident that the situation of such states as are unarmed or only partially equipped for defense would be extremely difficult, dangerous, and disadvantageous—a condition which is refuted by the facts.

I am convinced that demonstrative reference to armaments are but a weapon in the hands of the factors interested in their production or in the maintenance and development of a military system for financial or political—egotistic—reasons.

I am firmly of the opinion that the educational effect of a first and genuine achievement in the realm of disarmament would prove highly efficacious, because the succeeding second and third steps would then be immeasurably simpler than the initial one; this for the obvious reason that the first results of an understanding would considerably weaken the familiar argument for national security with which parliamentarians of all countries now permit themselves to be intimidated.

Armaments can never be viewed as an economic asset to a state. They must ever remain the unproductive exploitation of men and material and an encroachment on the economic reserves of a state through the temporary conscription of men in the active periods of their lives—not to mention the moral impairment resulting from a preoccupation with the profession of war and the moral processes of preparing a nation for it.

NOTES ON PACIFISM

A pacific settlement of conflicts and the international cooperation of intellectuals is not possible until military service and the armies are abolished. Men of standing would do a great service to humanity by endorsing the refusal of our young men to perform military service. I am of the opinion that all thinking men should take a solemn pledge never to participate in any military activity, directly or indirectly.

* * *

I am rarely enthusiastic about what the League of Nations has done or has not done, but I am always thankful that it exists.

* * *

I hold that mankind is approaching an era in which peace treaties will not only be recorded on paper, but will also become inscribed in the hearts of men.

* * *

Peace cannot be kept by force. It can only be achieved by understanding. You cannot subjugate a nation forcibly unless you wipe out every man, woman, and child. Unless you wish to use so drastic a measure, you must find a way of settling your disputes without resort to arms.

* * *

Internationalism does not mean the surrender of individuality. There is no reason why a nation or a race should not preserve its traditions. Why should the Jew ignore his past? Why should he fail to call some spot on the face of the globe his own?

I can see no wrong in enlightened patriotism, in love of country and of race. But patriotism is no excuse for any group of men to assail its neighbors or to impress its point of view upon others by fire and sword. I believe with President Wilson in self-determination for nations and for individuals.

THE JEWS

THE JEWISH HOMELAND

It is misleading to speak of the Christianization of the Jews or the Judaization of the Christians. What is happening is that the two groups are undergoing a common development on common ground and are both being influenced by the same factors, just as is the case with the heterogeneous mixture of peoples that inhabit Europe.

To me the real problem is this: Why do so many Europeans and Americans bother so much about the little handful of Jews? But we cannot answer this question; the finding of the proper solution must be left to the others.

The most deadly foes of Jewish national consciousness and Jewish national dignity are fatty degeneration—a lack of racial consciousness which is a result of excessive wealth and comfort—and a certain sort of inner dependence upon the non-Jewish environment, a dependence proceeding from the increased laxity of Jewish communal authority. The finest elements of the human soul can flourish only in the fertile soil of a community. How doubly great, therefore, is the moral danger of the Jew who has lost his kinship with his own group and whom the people of the nation in which he lives regard as an alien! Frequently such a situation has produced a harsh and dismal egoism.

The external pressure bearing upon the Jewish people is particularly great at present. But our wretched position has proved beneficial to us. It has given rise to a revival of Jewish community life of which the generation before the last would never have dreamed.

Thanks to this newly aroused feeling of solidarity among the Jews, the colonization of Palestine—which our devoted and far-sighted leaders have already brought well under way, despite apparently unconquerable difficulties—has achieved such excellent results that I have no doubt as to its permanent success. The value of this work for all the Jews of the world is very high. Palestine will be a cultural center for all of Jewry, a refuge for the persecuted, a field of activity for the best among us—an ideal that will unite all the Jews of all the world and bring them spiritual regeneration.

I have seen our youth in Palestine cheerfully doing hard labor which better and more modern equipment could have rendered less difficult and more productive. I have seen a supremely courageous little band of colonists weighted down under an enormous burden of debts which some of their enthusiastic Zionist brethren could so easily have lightened. These young people must be helped as much as possible in their struggle. Their physical well-being must be preserved. It is our sacred duty; for they are sacrificing themselves for the soul and the repute of the entire Jewish people.

We must prove that we are a people of sufficient vigor and vitality to accomplish our great task, to create a center and a support for future generations. May the land of Palestine come to mean to us and to our children what the Temple of Solomon meant to our ancestors!

No Jew can truly sense this meaning unless he joyfully contributes his share to the Jewish work, the Jewish cause.

With the vision of genius, Herzl saw that the rehabilitation of Palestine would lend a new cohesive force, a new meaning, and a new dignity to the Jewish people.

We must not merely aid in this work, but we must adhere to the lofty outlook of the founder of the movement.

In view of the present situation of world Jewry, it is now more then ever necessary to preserve the Jewish community in a vital form. This end can best be attained by the colonization of Palestine, a work in which world Jewry is united, and by the fostering of the Jewish spiritual tradition.

The publication of my book in the language of our fathers fills me with particular delight. It is symptomatic of the metamorphosis which that language has undergone. Its use is no longer limited to the communication of purely Jewish matters to Jews; it is beginning to embrace all fields of human interest. This revival of our tongue constitutes an important factor in our struggle for independence.

The rebuilding of Palestine as the Jewish National Home differs fundamentally from all other Jewish activities of our time. This is a movement to aid not individuals, but an entire nation. The Jewish people will have to provide funds for this constructive work for many years to come. So stupendous and unique a task as the upbuilding of Palestine must take a course of constant, gradual development.

But our work is progressing. In recent years large and valuable stretches of Palestinian land have become the property of the Jewish people. Jewish hands are reclaiming more and more neglected and waste lands and transforming them into fertile fields and orchards.

We hope that Jewish national life in Palestine will make sufficiently great strides to become the basis of a new intellectual and cultural creativeness. The Jewish people—free of petty chauvinism and of the evils of European nationalism, living peacefully side by side with the Arabs, who enjoy equal rights—should be enabled to lead its national life in its ancient homeland, so that it may again assume a dominant rôle in the civilization of the world. Situated as it is on the borderland between the Orient and the Occident, the Jewish National Home may be able to play an important part in the development of a new humanity.

Jewish nationalism is a necessity because only through a consolidation of our national life can we eliminate those conflicts from which the Jews suffer to-day. May the time soon come when this nationalism will have become so thoroughly a matter of course that it will no longer be necessary for us to give it special emphasis.

I believe in the actuality of Jewish nationality, and I believe that every Jew has duties toward his coreligionists. The meaning of Zionism is thus many-sided. It opens out to Jews who are despairing in the Ukrainian hell or in Poland, hope for a more human existence. Through the return of Jews to Palestine, and thus back to normal and healthy economic life, Zionism means, too, a productive function, which should enrich mankind at large. But the chief point is that Zionism must tend to strengthen the dignity and self-respect of Jews in Diaspora. I have always been annoyed by the undignified assimilationist cravings and strivings which I have observed in so many of my friends.

Through the founding of a Jewish Commonwealth in Palestine, the Jewish people will again be in a position to bring their creative faculties into full play. Through the erection of the Hebrew University and similar institutions, the Jewish people will not only help their own national renaissance, but will enrich their moral culture and knowledge; and, as in centuries past, be directed to new and better ways than those which present world conditions necessarily entail for them.

A special task devolves upon the Hebrew University in the spiritual direction and education of the laboring sections of our people in the land. In Palestine it is not our aim to create another people of city dwellers leading the same life as in the European cities and possessing the European bourgeois standards and conceptions. We aim at creating a people of workers, at creating the Jewish village in the first place, and we desire that the treasures of culture should be accessible to our laboring class, especially since Jews in all circumstances, as we know, place education above all things. In this connection it devolves upon the University to create something unique in order to serve the specific needs of the forms of life developed by our people in Palestine.

A University is a place where the universality of the human spirit manifests itself. Science and investigation recognize as their aim the truth only. It is natural, therefore, that institutions which serve the interests of science should be a factor making for the union of nations and men. Unfortunately, the universities of Europe to-day are for the most part the nurseries of chauvinism and of a blind intolerance of all things foreign to the particular nation or race, of all things bearing the stamp of a different individuality. Under this régime the Jews are the principal sufferers, not only because they are thwarted in their desire for free participation and in their striving for education, but also because most Jews find themselves particularly cramped in this spirit of narrow nationalism. I should like to express the hope that our University will always be free from this evil, that teachers and students will always preserve the consciousness that they serve their people best when they maintain its union with humanity and with the highest human values.

ADDRESS

DELIVERED IN LONDON, OCTOBER 27, 1930, BEFORE THE ORT AND OZE SOCIETIES *

It is no easy task for me to overcome my inclination to a life of quiet contemplation. Nevertheless, to the cry of the Ort and Oze Societies I have been unable to turn a deaf ear, for it is at the same time, as it were, the cry of our heavily burdened people, to whose voice I respond.

The situation of our Jewish communities

* The Ort and Oze Societies are international organizations working for the betterment of living and working conditions among the Jews of eastern Europe.

scattered throughout the world forms at once a barometer of the moral standard in the political world. For what could be more characteristic of the level of political morality and righteousness than the attitude of the nations toward a defenseless minority whose peculiarity it is to preserve its ancient traditions of culture?

In our day this barometer stands very low. We feel it painfully in our fate. But even this very depression confirms me in the conviction that the preservation and the consolidation of this community is our duty. Within the traditions of the Jewish people exists a striving toward righteousness and understanding that should be of service to the rest of the nations, both now and in the future. Spinoza and Karl Marx are the children of this tradition.

Whoever will preserve the spirit must also take care of the body to which the spirit is bound. The Oze Society literally cares for the body of our people in eastern Europe. It is working indefatigably for the physical preservation of our economically heavily burdened people, while the Ort Society is striving to remove a social and economically burdensome wrong from which the Jewish people have suffered from the time of the Middle Ages. Because in the Middle Ages all the directly productive vocations were closed to us, we were driven to adopt purely mercantile vocations.

Address to Ort and Oze Societies

The only effective help that can be given the Jewish people in these eastern lands is to throw open to them the new fields of vocational activity for which they are striving all over the world. This is the difficult problem which the Ort Society is successfully tackling.

To you, our English brethren, has now come the call to collaborate in the great work which has been inaugurated by celebrated men. The last years, yes, even the last days, have brought us a disappointment which you in particular must feel keenly. Do not bemoan the hardness of fate, but in this occurrence see rather a motive for both being and remaining faithful to the Jewish community. I am firmly convinced that thus we shall serve indirectly the aims of humanity in general, which aims must forever remain, with us, the highest.

Remember that difficulties and obstacles always form for any community a valuable source of strength and health. We should not have survived as a community all the centuries if we had had a bed of roses. Of that I am strongly convinced.

But we have a still better consolation. The number of our friends is not great, but among them are men of lofty spirit and righteousness who have devoted their lives to ennobling the human race and to the liberation of individuals from degrading oppression.

Address to Ort and Oze Societies

We are glad and happy that we have with us to-day such men from the non-Jewish world; men who, through this memorable evening, lend a special dignity and solemnity. It rejoices me to see before me Bernard Shaw and H. G. Wells, for whose conception of life I have a special feeling of sympathy.

OPINIONS AND APHORISMS

ON RADIO

One ought to be ashamed to make use of the wonders of science embodied in a radio set, the while appreciating them as little as a cow appreciates the botanic marvels in the plants she munches.

Let us not forget how humanity came into possession of this wonderful means of communication. The source of all scientific advancement is the God-given curiosity of the toiling experimenter and the constructive fantasy of the technical inventor.

Remember Oerstedt, who first discovered the magnetic influence of electro-magnetic currents; remember Reis, who first employed this influence to create sound in an electro-magnetic way; Bell, who, by using sensitive contacts, transferred sound waves with his microphone into variable electric currents. Remember furthermore, Maxwell, who mathematically proved the existence of electric waves, and Hertz, who first created them with the help of a spark. Think especially of Lieben, who, with his Fleming valve, invented an incomparable detector organ for electric waves which simultaneously turned out to be an ideally simple instrument for the creation of electric waves. Remember thankfully the army of nameless technicians who simplified radio instruments and adapted them to mass production so that they became accessible to everybody.

It was the scientists who first made true democracy possible, for not only did they lighten our daily tasks but they made the finest works of art and thought, whose enjoyment until recently was the privilege of the favored classes, accessible to all. Thus they awakened the nations from their sluggish dullness.

The radio broadcast has a unique function to fill in bringing nations together. It can be used for strengthening that feeling of mutual friendship which so easily turns into mistrust and enmity.

Until our day people learned to know each other only through the distorting mirror of their own daily press. Radio shows them to each other in the liveliest form, and, in the main, from their most lovable sides.

ON SCIENCE

I believe in intuition and inspiration. . . . At times I feel certain I am right while not knowing the reason. When the eclipse of 1919 confirmed my intuition, I was not in the least surprised. In fact, I would have been astonished had it turned out otherwise. Imagination is more important than knowledge. For knowledge is limited, whereas imagination embraces the entire world, stimulating progress, giving birth to evolution. It is, strictly speaking, a real factor in scientific research.

* * *

Cosmic Religion

The basis of all scientific work is the conviction that the world is an ordered and comprehensive entity, which is a religious sentiment. My religious feeling is a humble amazement at the order revealed in the small patch of reality to which our feeble intelligence is equal.

* * *

By furthering logical thought and a logical attitude, science can diminish the amount of superstition in the world. There is no doubt that all but the crudest scientific work is based on a firm belief—akin to religious feeling—in the rationality and comprehensibility of the world.

* * *

Music and physical research work originate in different sources, but they are interrelated through their common aim, which is the desire to express the unknown. Their reactions are different, but their results are supplementary. As to artistic and scientific creation, I hold with Schopenhauer that their strongest motive is the desire to leave behind the rawness and monotony of everyday life, so as to take refuge in a world crowded with the images of our own creation. This world may consist of musical notes as well as of mathematical rules. We try to compose a comprehensive picture of the world in which we are at home and which gives us a stability that cannot be found in our external life.

* * *

Science exists for Science's sake, like Art for Art's sake, and does not go in for special pleading or for the demonstration of absurdities.

* * *

A law cannot be definite for the one reason that the conceptions with which we formulate it develop and may prove insufficient in the future. There remains at the bottom of every thesis and of every proof some remainder of the dogma of infallibility.

* * *

In every naturalist there must be a kind of religious feeling; for he cannot imagine that the connections into which he sees have been thought of by him for the first time.

He rather has the feeling of a child, over whom a grown-up person rules.

* * *

We can only see the universe by the impressions of our senses reflecting indirectly the things of reality.

* * *

Among scientists in search of truth wars do not count.

* * *

There is no universe beyond the universe for us. It is not part of our concept. Of course, you must not take the comparison with the globe literally. I am only speaking in symbols. Most mistakes in philosophy and logic occur because the human mind is apt to take the symbol for the reality.

* * *

Cosmic Religion

I see a pattern. But my imagination cannot picture the maker of that pattern. I see the clock. But I cannot envisage the clockmaker. The human mind is unable to conceive of the four dimensions. How can it conceive of a God, before whom a thousand years and a thousand dimensions are as one?

* * *

Imagine a bedbug completely flattened out, living on the surface of a globe. This bedbug may be gifted with analysis, he may study physics, he may even write a book. His universe will be two-dimensional. He may even intellectually or mathematically conceive of a third dimension, but he cannot visualize it. Man is in the same position as the unfortunate bedbug, except that he is three-dimensional. Man can imagine a fourth dimension mathematically, but he cannot see it, he cannot visualize it, he cannot represent it physically. It exists only mathematically for him. The mind cannot grasp it.

MISCELLANEOUS

Everyone sits in the prison of his own ideas; he must burst it open, and that in his youth, and so try to test his ideas on reality. But in a couple of centuries there comes another, perhaps, who refutes him. It is true that this will not happen to the artist in his uniqueness. It is all within the nature of research and it is not at all sad.

* * *

Youth is always the same, endlessly the same.

* * *

I do not believe individuals possess any

unique gifts. I only believe that there exists on one hand talent and on the other hand developed qualifications.

* * *

In Mme. Curie I can see no more than a brilliant exception. Even if there were more woman scientists of like caliber they would serve as no argument against the fundamental weakness of the feminine organization.

* * *

Before God we are relatively all equally wise—equally foolish.

* * *

Working is thinking, hence it is not always easy to give an exact accounting of one's time. Usually I work about four to six hours a day. I am not a very diligent man.

* * *

The intellectuals always have microscopes before their eyes.

* * *

Never forget that the fruit of our labor does not constitute an end in itself. Economic production should make life possible, beautiful, and noble. We must not permit ourselves to be degraded into mere slaves of production.

* * *

Hitler is no more representative of the Germany of this decade than are the smaller anti-Semitic disturbances. Hitler is living—or shall I say sitting?—on the empty stomach of Germany. As soon as economic conditions improve, Hitler will sink into oblivion. He dramatizes impossible extremes in an amateurish manner.

Reduced to a formula, one might say simply that an empty stomach is not a good political adviser. Unfortunately, the corollary also is true, namely, that better political insight has a hard time winning its way as long as there is little prospect of filling the stomach.

Personally, I feel that there is enough technical knowledge accumulated in the world to-day to make conditions such as we have in Germany unnecessary. It should be possible to produce enough of the necessaries of life to satisfy everybody and at the same time give work to everybody. That, of course, means short hours and high wages, and not, as is so often advocated, longer hours and lower wages.

* * *

Mass psychology is a difficult thing to fathom. I fear historians never have taken the factor of mass psychology sufficiently into account in writing history. They look upon events in retrospect with the idea that they can define exactly the causes that led up to this or that outstanding event. In reality, behind these apparent causes there are indefinable factors of mass psychology about which we know little or nothing.

My own case is, alas, an illustration. Why popular fancy should seize upon me, a scientist dealing in abstract things and happy if left alone, is one of those manifestations of mass psychology that are beyond me. I think it is terrible that this should be so and I suffer more than anybody can imagine.

* * *

I dislike to apply a yardstick to such imponderables as genius. Shaw is undoubtedly one of the world's greatest figures, both as a writer and as a man. I once said of him that his plays remind me of Mozart.

There is not one superfluous word in Shaw's prose, just as there is not one superfluous note in Mozart's music. The one in the medium of language, the other in the medium of melody, expresses perfectly with almost superhuman precision, the message of his art and his soul.